ISAAC ASIMOV'S NEW LIBRARY OF THE UNIVERSE
SPACE COLONIES

BY ISAAC ASIMOV
WITH REVISIONS AND UPDATING BY GREG WALZ-CHOJNACKI

Gareth Stevens Publishing
MILWAUKEE

For a free color catalog describing Gareth Stevens' list of high-quality books, call 1-800-542-2595 (USA) or 1-800-461-9120 (Canada). Gareth Stevens' Fax: (414) 225-0377.

The reproduction rights to all photographs and illustrations in this book are controlled by the individuals or institutions credited on page 32 and may not be reproduced without their permission.

Library of Congress Cataloging-in-Publication Data

Asimov, Isaac.
   Space colonies / by Isaac Asimov; with revisions and updating by
  Greg Walz-Chojnacki.
     p. cm. — (Isaac Asimov's New library of the universe)
   Rev. ed. of: Colonizing the Planets and Stars. 1990.
   Includes index.
   ISBN 0-8368-1225-5
    1. Space colonies—Juvenile literature. [1. Space colonies.]
  I. Walz-Chojnacki, Greg, 1954-. II. Asimov, Isaac. Colonizing
  the planets and stars. III. Title. IV. Series: Asimov, Isaac.
  New library of the universe.
  TL795.7.A852   1995
  919.9'04--dc20   95-7229

This edition first published in 1995 by
Gareth Stevens Publishing
1555 North RiverCenter Drive, Suite 201
Milwaukee, Wisconsin 53212, USA

Revised and updated edition © 1995 by Gareth Stevens, Inc.
Original edition published in 1990 by Gareth Stevens, Inc. under the title
Colonizing the Planets and Stars. Text © 1995 by Nightfall, Inc.
End matter and revisions © 1995 by Gareth Stevens, Inc.

All rights to this edition reserved to Gareth Stevens, Inc. No part of this book may be reproduced, stored in a retrieval system, or transmitted in any form or by any means, electronic, mechanical, photocopying, recording, or otherwise without the prior written permission of the publisher except for the inclusion of brief quotations in an acknowledged review.

Series editor: Barbara J. Behm
Design adaptation: Helene Feider
Production director: Teresa Mahsem
Editorial assistant: Diane Laska
Picture research: Matthew Groshek and Diane Laska

Printed in the United States of America

1 2 3 4 5 6 7 8 9 99 98 97 96 95

To bring this classic of young people's information up to date, the editors at Gareth Stevens Publishing have selected two noted science authors, Greg Walz-Chojnacki and Francis Reddy. Walz-Chojnacki and Reddy coauthored the recent book *Celestial Delights: The Best Astronomical Events Through 2001.*

Walz-Chojnacki is also the author of the book *Comet: The Story Behind Halley's Comet* and various articles about the space program. He was an editor of *Odyssey,* an astronomy and space technology magazine for young people, for eleven years.

Reddy is the author of nine books, including *Halley's Comet, Children's Atlas of the Universe, Children's Atlas of Earth Through Time,* and *Children's Atlas of Native Americans,* plus numerous articles. He was an editor of *Astronomy* magazine for several years.

CONTENTS

An Endless Migration	4
Saving Our Species	7
Onward to the Moon and Mars	9
The Asteroid People	10
Life in Space	12
Making New Earths	15
Colonizing Our Solar System	16
Beyond Our Solar System	19
More Power to Them	20
Traveling Without Leaving Home	23
Star Cruising	24
No Longer Home	27
Fact File: Settling the Cosmos	28
More Books about Space Colonies	30
Videos	30
Places to Visit	30
Places to Write	30
Glossary	31
Index	32

We live in an enormously large place – the Universe. It's just in the last fifty-five years or so that we've found out how large it probably is. It's only natural that we would want to understand the place in which we live, so scientists have developed instruments – such as radio telescopes, satellites, probes, and many more — that have told us far more about the Universe than could possibly be imagined.

We have seen planets up close. We have learned about quasars and pulsars, black holes, and supernovas. We have gathered amazing data about how the Universe may have come into being and how it may end. Nothing could be more astonishing.

But now, it does not seem enough to simply observe the Universe. A more active role must be taken in dealing with the worlds beyond our own. That is because in the next century, Earthlings will probably live in various parts of the Universe. What will our new homes be like? How will we manage in space? Indeed, why should this quest be undertaken at all?

Isaac Asimov

An Endless Migration

Most scientists agree that the first human ancestors appeared in Africa about two million years ago. Our ancestors gradually migrated, and now there are almost six billion humans living nearly everywhere on the planet. The urge to move outward is still with us after all this time – and today it extends beyond Earth, into space.

Why do we dream of reaching out to the cosmos? There are many reasons. Many people worry that one day we will use up Earth's natural resources, making it necessary to move onto other worlds. Many people feel that chemicals we release into the atmosphere will destroy Earth's protective ozone layer, making Earth a hostile world. Others simply feel that humans will always have the need to settle new regions, even if it means going into space.

No other known place suits humans the way Earth does. But perhaps other worlds can be made livable for humans. Perhaps, one way or another, human beings can thrive somewhere beyond Earth.

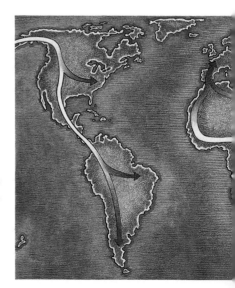

Opposite: Our earliest ancestors gazed at the Moon and stars with fear and wonder.

Inset: The oldest human fossils have appeared in central Africa. From there, our ancestors slowly migrated to new lands all across the globe. Perhaps one day our migration will take us to worlds beyond Earth.

Saving Our Species

Our planet Earth will not be a suitable home forever. For instance, comets or asteroids could strike Earth causing loss of life. The dinosaurs became extinct in this manner. If humans had been alive at that time, they probably would have been wiped out, too. To avoid future catastrophe, it is important to have a plan of action should a comet or asteroid strike Earth again.

But even if we learn to protect ourselves from this kind of situation, scientists know that the star that gives us life – our Sun – won't last forever. Eventually, it will swell up and roast our rocky home, then fade away into a dim white cinder. When that happens, Earth will not be fit for life of any kind.

The *good* news is that we have plenty of time to prepare for that – the Sun should shine steadily for at least a couple of billion more years.

Inset: Another comet or asteroid strike like the one that led to the extinction of the dinosaurs could mean trouble for us, as well. We might be clever enough to survive, but it would be wise to have a "summer home" on another planet, just in case.

Opposite: As our Sun dies, it will first swell into a "red giant," greatly raising the temperature on Earth and making life impossible.

Onward to the Moon and Mars

As humans begin inhabiting other worlds, the Moon will be needed as a mining base. The Moon's soil can be used to make metals, glass, oxygen, and other substances necessary to build settlements and structures in space. Cities can also be built underground on the Moon with resources, such as water, brought from Earth.

Mars can also be inhabited by humans. It is bigger than the Moon and has its own water. It, too, can have underground cities. Humans on the Moon and Mars will have to become used to living inside structures.

Top: An artist's idea of an outpost on Mars.

Bottom, left: Four American scientists stayed in this underwater habitat for three months, proving that humans can survive in alternative environments.

Bottom, right: Water is one ingredient necessary for life as we know it. In this illustration, water vapor condenses and eventually returns to Earth as rain or snow.

! *Worlds in balance*

Humans won't be alone on other worlds. Plants (for food) will have to be brought along – and animals, too. The animals will need different foods from those that humans eat. For example, if birds are brought along to the new worlds, they will need insects to eat. A good balance of Earth's various elements will be necessary on new worlds. Scientists will need to study each new situation carefully to find the best possible plan for settlements in space.

The Asteroid People

Beyond Mars lies the asteroid belt. It contains perhaps as many as 100,000 asteroids orbiting the Sun between Mars and Jupiter. Asteroids might be a wonderful resource for cosmic colonizers of the future. For one thing, asteroids could be the largest source of minerals anywhere in the Solar System. In addition, even the small ones could be hollowed out to make worlds larger than any we can build in space for ourselves.

Someday there may be many thousands of settlements in the asteroid belt, each with a million people or more, each with its own customs and culture. There might be as many people in the asteroids as on Earth, and it may be the "asteroid people" who will make the long trips to the outer Solar System.

Opposite: In this drawing, future colonists have created a spaceship from a hollowed-out asteroid.

Top: In this illustration, astronauts are on board an asteroid that is headed back to Earth. There, the asteroid will be mined for useful materials.

Bottom: The human body has adapted to the daily cycle of Earth's rotation. In one famous early experiment, scientist Stefania Follini spent over 130 days alone deep within a cave. When she emerged, her body had formed new rhythms. What will happen to humans who spend years in space?

❓ Questions without answers

Will everything go smoothly for people when they travel to a space colony, leaving Earth forever? Who will go? Who will stay behind? People with certain talents and skills will undoubtedly go. Will they be homesick for our Solar System? Will they be frightened by the vast emptiness between the stars? Must everyone be alike? Will it be possible to build a healthy starship society?

Life in Space

With people living in space, space travel will become quite common. Many spaceships will be built while in space and will coast easily from one place to another because they won't have to fight Earth's gravity to do so. There will be ships carrying goods from one settlement to another, rescue ships in case of accidents, and repair ships to keep space settlements and underground cities in good condition.

Many of these ships could be run by robots equipped to do simple jobs. Robots could also run factories in space and work at solar power stations that extract energy from the Sun.

Opposite: Robot helpers assist astronauts in the construction of a space habitat.

Below, left: A robotic space ambulance.

Below, right: In past missions, spaceships called *Progress* ships from the former Soviet Union brought fresh supplies to orbiting cosmonauts. The unpiloted vehicles were controlled from the ground.

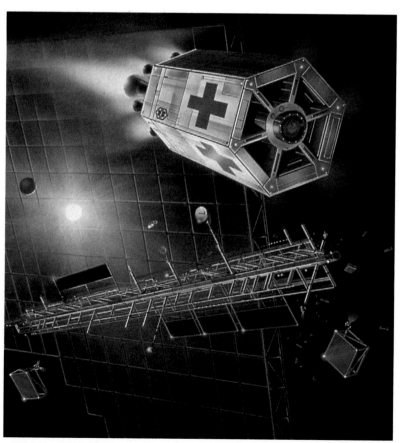
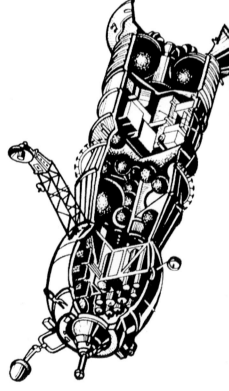

Making New Earths

Scientists are already learning how to make safe living spaces for humans in the hostile conditions on other planets. Some scientists think it would even be possible to change a planet like Mars into a new version of Earth. Changing a planet to make it Earthlike is called "terraforming."

Mars is a cold world with an atmosphere that's too thin to breathe. This atmosphere also allows dangerous ultraviolet radiation from the Sun to reach the surface of the planet.

By collecting extra sunshine, scientists could slowly warm Mars, freeing the frozen water that lies in the icecaps and below the ground. Plants could make breathing in the planet's atmosphere possible. The transformed atmosphere would also protect inhabitants from the Sun's dangerous rays.

Terraforming a planet like Mars or Venus would take hundreds or thousands of years. With careful planning, though, "Gardens of Eden" could be created.

Inset: This illustration shows what Mars might be like during the process of terraforming. Humans would no longer need space suits. They would, however, need to carry some extra oxygen when participating in activities such as looking for plant life.

Opposite: Scientists are certain Mars contains plenty of frozen water.

Colonizing Our Solar System

The outer planets lie far beyond the asteroid belt. It would take a very long time to reach the ends of the Solar System from Earth as well as from the asteroids. But the asteroid people could make the trip manageable by building large ships with crews of hundreds. Such ships could be little worlds of their own so the crew would feel comfortable and at home during the long trip.

We might be able to settle some of the moons of the outer planets, but the gas giants themselves – Jupiter, Saturn, Uranus, and Neptune – are far too hostile for human use. Eventually, we may even build an outpost on distant Pluto, where the Sun would seem only like a bright, distant star. There, from the rim of our Solar System, people could look outward toward the other stars.

Opposite: Imagine growing up with a huge planet like Jupiter in your sky, instead of Earth's small Moon.

Inset: A starship approaches one of the worlds in its new solar system.

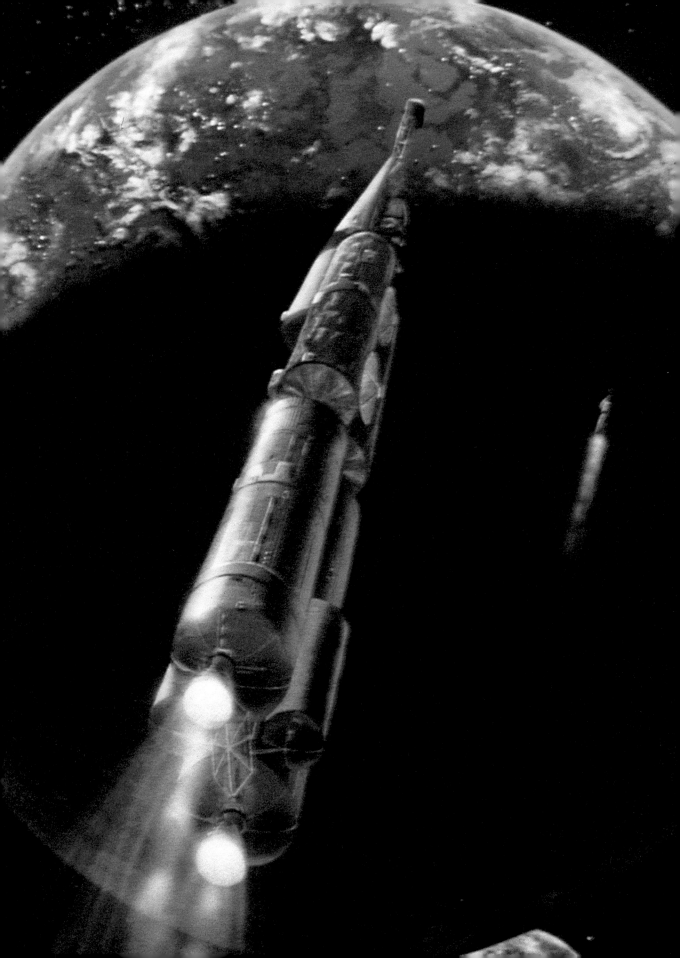

Beyond Our Solar System

What happens when our Solar System is completely settled? Where do we go next? There are trillions of other stars, many of which must have planets orbiting them. But can we reach them? Even the nearest star is about seven thousand times as far away from us as Pluto.

The speed of light is about 186,000 miles (300,000 kilometers) per second. That seems very speedy – faster than anything we can imagine here on Earth. But even at the speed of light, it would take over four years to reach Alpha Centauri, Earth's nearest star, from Earth. It would take even longer to reach other stars, and 100,000 light-years to travel from one end of our Milky Way Galaxy to the other.

! *Other galaxies – so "near" and yet so far*

Imagine that we've inhabited every star system in our Galaxy that might have planets fit for humans. Must we stop then? Not necessarily. There are other galaxies. For instance, three small galaxies, the Magellanic Clouds, are over 150,000 light-years away.

The mammoth Andromeda Galaxy is over two million light-years away. It would take millions of years to reach these galaxies from Earth, but perhaps we can do it if our descendants build starships capable of drifting long enough.

Left: The distance between stars in our Milky Way Galaxy will keep us in the neighborhood of our Sun for a long time. Perhaps one day, though, we will want to take another giant leap – to a nearby galaxy. The nearest galaxy like our own, the Andromeda Galaxy, is so far away that its light takes over two million years to reach us.

Opposite: A starship blasts out of Earth's orbit en route to the stars.

More Power to Them

How would scientists build up enough energy in space vehicles for the vehicles to fly even to the closest stars? Instead of using rocket fuel, objects called ion drives could be used to push tiny charged atoms – ions – backward. This would create speed more efficiently than today's engines.

Spaceships could also be powered with laser beams. But no matter how much energy is used, it will still take many years to reach even the nearby stars.

Opposite, bottom: The Orion Nuclear-pulse *(in this drawing)* was an idea popular in the 1950s. Had it become a reality, nuclear explosions would have propelled spacecraft.

Above: The Bussard Ramjet Starship is depicted by an artist. The craft would "scoop" hydrogen out of space to be used as fuel.

❗ A galaxy of wonders

Many cosmic oddities await curious star travelers. Imagine encountering a white dwarf star, an incredibly dense object but smaller than Earth. A neutron star is also massive but less than 10 miles (16 km) across. Its gravitational pull is so great you wouldn't want to come too close. Imagine encountering a supergiant, a star that is hundreds of millions of miles (km) wide. The Universe is full of unusual and marvelous sights.

Traveling without Leaving Home

Of course, space settlers might not care about moving quickly through the cosmos. They might want to take their time. An entire asteroid settlement might decide to become a "starship" and leave the Solar System. They could outfit their settlement with advanced ion drives and take off, coasting at speeds of mere hundreds of miles (km) a second.

Many generations would be born and would die before they reached another world, but that wouldn't matter. They wouldn't have left home. Home would have come right along with them.

Opposite: Mobile living units on an asteroid.

Inset: A mobile living unit on Earth.

? *A galactic baby boom?*

A starship in flight can only support so many people, so it must control its population. Once another star is sighted with a planet like Earth, or with an asteroid belt, things would be different. People could settle the planet or build asteroid homes. They could then safely increase their population. When, after thousands of years, the new worlds are full, another group of starships might set out. Eventually, the entire Galaxy might be settled in this way.

Star Cruising

Let's imagine that such starships become common and slowly spread out from our Solar System in every direction.

These ships might want to stay in touch with each other, and with Earth. There might be special "star crews" that would travel quickly from Earth to a spaceship to deliver messages, passengers, and other items, and then return again to Earth.

Travelers on Earth have "jet lag" when they change time zones. Think of the incredible adjustments to a new time period star crews would have to make.

Opposite: During long trips to other stars, crews of starships would have to entertain themselves. Here crew members play a cosmic game.

Right: What unusual worlds and alien skies will become home to future generations of humans? While such sights might seem strange to us, our Earth will seem as strange to our descendants who return to Earth generations from now.

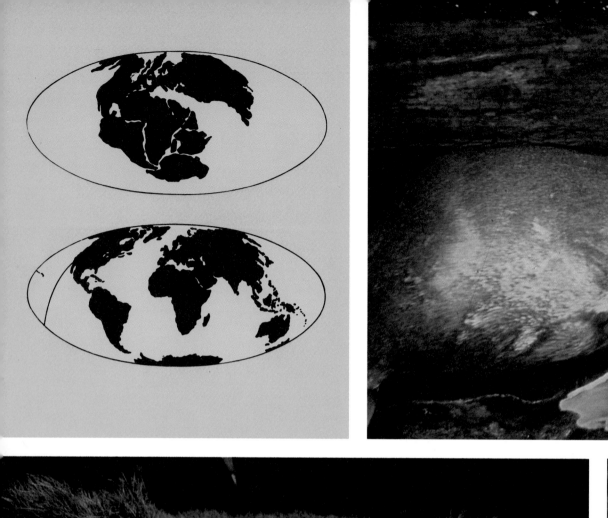

No Longer Home

As humanity spreads out through the Galaxy over millions of years, people in one star system might know nothing about people in others. They may all evolve in different directions, populating the Galaxy with millions of different kinds of people.

Ten million years from now, humans from space might come across Earth while they are out exploring. Earth will have changed tremendously. Will the explorers from space recognize their homeland? Perhaps not.

❓ Are we alone?

Can we assume people are the only living things in the Galaxy with our kind of intelligence? It is hard to believe there aren't intelligent beings elsewhere among all the hundreds of billions of stars in the Galaxy. Will our starships reach worlds with intelligent beings on them? What if we encounter starships with nonhumans? We might find them so different that we can't relate to them. Or we might find we can learn a great deal from them.

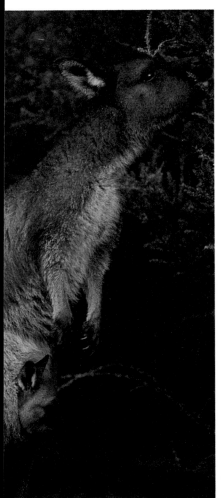

Opposite: Two hundred million years ago as the continents of Earth slowly separated *(top, left)*, isolated land areas like Australia developed unique forms of life. The kangaroo *(bottom, right)*, koala *(bottom, left)*, and duck-billed platypus *(top, right)* exist in the wild only in Australia. As humans expand into the Galaxy, each colony will develop differently. When descendants of different colonies meet millions of years in the future, they may not even realize they are related.

Fact File: Settling the Cosmos

Is your curiosity cosmic? If so, the idea of settling planets and distant star systems might appeal to you. The chart on these pages gives you an idea of just a few of the things you would need to consider before signing up to settle space.

Before attempting to settle the Moon, Mars, and interstellar space, Earthlings may want to set up Earth-orbiting colonies close to home. Like space stations, these colonies would give people a place to live and work just beyond Earth's atmosphere. These colonies would house research labs, relay stations, and even factories in space. Beyond the Moon and Mars, and before taking the leap into deep space, we might want to settle the asteroid belt or some other part of the outer Solar System.

Colonizing the Moon, Mars, and Beyond — Some Tricks of the Trade

Destination	Travel Time	Gravity
Earth's Moon	One-way trip from Earth: several days. Communications with Earth: 3 seconds for round-trip transmission.	One-sixth that of Earth's, so daily exercise routine necessary.
Mars	One-way trip from Earth: 6-9 months. Communications with Earth: 10-40 minutes for round-trip transmission.	Two-fifths that of Earth's. Exercise regimen a must.
The Stars	One-way trip from Earth: indefinite. Closest star, Alpha Centauri, is "only" 4.2 light-years away, but will we ever be able to travel that fast? Trip might have to be made in stages over many generations. Communications with Earth: years; eventually, none likely.	Artificial gravity possible by rotating deep-space colony. Also possible to adjust to gravity of final destination. Travelers will have years, possibly generations, to adapt to "new" gravity.

As with the Moon or Mars, mining might be a good reason to set up shop among the asteroids. Or we might want to settle the asteroids to relieve overcrowding and the taxing of natural resources on Earth.

All this sounds exciting, but it would also be a huge challenge. Think of creating an artificial atmosphere; harnessing solar, nuclear, and other forms of energy; producing "natural" resources in space; and controlling the environment in general over an incredibly long time. The demands of time and energy are great. But if we move step by step, starting with Earth-orbiting colonies and lunar colonies, both time and science could be on our side to propel us toward the stars.

Needs	**Nonhuman Inhabitants**	**Civilization in Space**
Medical: Ferrying of medicine from Earth should be fairly routine. Colony will need its own doctors, nurses, and other medical staff. Other: Within a few days' travel time, supplies and other help are easily available from Earth.	Plants, animals, insects, and useful bacteria for farming and recycling air in permanent colony. Closeness to Earth would make experimenting and bringing in new varieties easy.	Because of proximity to Earth and ease of communication and travel between Earth and Moon, early settlers would probably be tied to Earth cultures from which they came. In time, as on Earth, settlements might increasingly take on their own character, along with their own distinct cultural identity.
Medical: Because of greater distance from Earth, any medical help that can't wait 6-9 months should be in-colony. Dentists now more crucial as low gravity can cause bone deterioration. Other: Long waits mark Earth-to-Mars shipping. Nearby space colonies likely drop-off points.	As on Moon, plants, animals, insects, and helpful bacteria for farming, breeding, and recycling. One big difference: Life-forms would be harder to replace because of distance from Earth.	As in Moon settlements, Mars colonists would be tied to Earth by fairly quick communications. But travel time and proximity to the asteroids and beyond might isolate settlers. Over time, it's possible that "Martian" communities would develop their own identity as a society and form their own governments.
Medical: Unlikely that deep-space settlements will have contact with – or even knowledge of – each other. Thus, all medical help must come from within each colony. Other: Everything must come from within each colony. Even "quickest" communication would take years.	There's no going back! Thus, any mistakes in combinations of plants, animals, insects, or bacteria could be deadly to the community. Must be sure of the right combination from the start.	Schools, medical books, tools, music, computers, factories, research labs, libraries, leisure activities and equipment, art, hospitals – all this and more, or the resources to develop these things – must be brought along. But after generations in space, the colony would evolve into a civilization unlike any known on Earth.

More Books about Space Colonies

Asimov on Astronomy. Asimov (Doubleday)
Colonies in Orbit. Knight (Morrow)
The High Frontier. O'Neill (Morrow)
Out of the Cradle: Exploring the Frontiers Beyond Earth. Hartmann (Workman)
Space Explorers. Asimov (Gareth Stevens)

Videos

The Asteroids. (Gareth Stevens)
Astronomy 101: A Beginner's Guide to the Night Sky. (Mazon)
Our Milky Way and Other Galaxies. (Gareth Stevens)

Places to Visit

You can explore the Universe – including the places where colonies may be established beyond Earth – without leaving our planet. Here are some museums and centers where you can find a variety of space exhibits.

Wings Over the Rockies
   Aviation and Space Museum
Hangar 1-Lowry Air Force Base
Denver, CO 80230-5000

National Air and Space Museum
Smithsonian Institution
Seventh and Independence Avenue SW
Washington, D.C. 20560

Anglo-Australian Observatory
Private Bag
Coonarbariban 2357 Australia

The Space and Rocket Center
   and Space Camp
One Tranquility Base
Huntsville, AL 35807

San Diego Aero-Space Museum
2001 Pan American Plaza
Balboa Park
San Diego, CA 92101

Seneca College Planetarium
1750 Finch Avenue East
North York, Ontario M2J 2X5

Places to Write

Here are some places you can write for more information about venturing into space. Be sure to state what kind of information you would like. Include your full name and address so they can write back to you.

National Space Society
922 Pennsylvania Avenue SE
Washington, D.C. 20003

Canadian Space Agency
Communications Department
6767 Route de L'Aeroport
Saint Hubert, Quebec J3Y 8Y9

Sydney Observatory
P.O. Box K346
Haymarket 2000 Australia

NASA Lewis Research Center
Educational Services Office
21000 Brookpark Road
Cleveland, OH 44135

Glossary

alien: in this book, a being from some place other than Earth.

asteroid: very small "planets" and even smaller objects made of rock or metal. There are thousands of asteroids in our Solar System, and they mainly orbit the Sun in large numbers between Mars and Jupiter. But some show up elsewhere in our Solar System – some as meteoroids and some possibly as "captured" moons of planets such as Mars.

asteroid belt: the space between Mars and Jupiter that contains thousands of asteroids.

atmosphere: the gases that surround a planet, star, or moon.

bacteria: the smallest and simplest forms of cell life. A bacterium is one-celled and can live in soil, water, air, food, plants, and animals, including humans.

billion: the number represented by 1 followed by nine zeroes – 1,000,000,000. In some countries, this number is called "a thousand million." In these countries, one billion would then be represented by 1 followed by twelve zeroes – 1,000,000,000,000 – a million million.

evolve: to develop or change over a long period of time.

generation: the average period of time between the birth of parents and the birth of their children.

gravity: the force that causes objects like the Earth and Moon to be attracted to one another.

ion drive: an engine that works by the ejection of charged atoms or molecules.

laser: L(ight) A(mplification) by S(timulated) E(mission) of R(adiation). A device that focuses light to a beam intense enough to burn holes through the hardest metals known.

Magellanic Clouds: the galaxies nearest the Milky Way, irregular in form and visible to the naked eye in the Southern Hemisphere.

Milky Way: the name of our Galaxy.

natural resources: valuable materials supplied by the environment.

neutron star: a star with all the mass of an ordinary large star but with its mass squeezed into a much smaller ball.

oxygen: the gas in Earth's atmosphere that makes human and other animal life possible. Simple life-forms changed carbon dioxide to oxygen as life evolved on Earth.

ozone layer: that part of our atmosphere that shields Earth from the Sun's dangerous ultraviolet rays.

Pluto: the farthest known planet in our Solar System and one so small that some scientists believe it to be a large asteroid.

red giant stars: huge stars that may be over 100 million miles (160 million km) in diameter.

Solar System: our Sun with the planets and all other bodies, such as the asteroids, that orbit the Sun.

terraforming: making another world like Earth by providing it with substances that are, as far as we know, special to Earth, such as oxygen and water.

white dwarf: the small white-hot body that remains when a star like our Sun collapses.

Index

Africa 4-5
Alpha Centauri 19, 28
Andromeda Galaxy 19
asteroids 6-7, 10-11, 16, 22-23, 28-29
atmosphere, Earth's 4, 29
atoms 20
Australia 26-27

Bussard Ramjet Starship 20-21

comets 6-7
cosmonauts 12

dinosaurs 6-7

Follini, Stefania 10

gravity 12, 20, 28

hydrogen 20-21

ion drives 20, 23
ions 20

Jupiter 10, 16-17

laser power 20

Magellanic Clouds 19
Mars 8-9, 10, 14-15, 28-29
Milky Way Galaxy 19, 23, 26-27

Moon, Earth's 4-5, 9, 16-17, 28-29

natural resources 4, 29
Neptune 16
neutron stars 20
nuclear power 20-21, 29

Orion Nuclear-pulse craft 20-21
ozone layer 4

Pluto 16, 19
Progress (spacecraft) 12

red giants 6-7
robots 12-13

Saturn 16
Solar System 10, 16, 19, 23, 24, 28-29
Sun 6-7, 10, 12, 15, 16, 19
supergiant stars 20

terraforming 14-15

ultraviolet radiation 15
Uranus 16

Venus 15

white dwarf stars 20

Born in 1920, Isaac Asimov came to the United States as a young boy from his native Russia. As a young man, he was a student of biochemistry. In time, he became one of the most productive writers the world has ever known. His books cover a spectrum of topics, including science, history, language theory, fantasy, and science fiction. His brilliant imagination gained him the respect and admiration of adults and children alike. Sadly, Isaac Asimov died shortly after the publication of the first edition of *Isaac Asimov's Library of the Universe*.

The publishers wish to thank the following for permission to reproduce copyright material: front cover, © Père Castor Flammarion; 4-5, 5, © Keith Ward 1989; 6, © Michael Carroll; 6-7, © Julian Baum; 8, official US Navy Photograph; 8-9 (upper), NASA, artwork by Pat Rawlings; 8-9 (lower), © Gareth Stevens, Inc. 1989; 10 (upper), © Mark Maxwell 1986; 10 (lower), © Mark Williams, Current Argus Photo, 1989; 11, © David Hardy; 12 (left), © Rick Sternbach; 12 (right), James Oberg Archives; 13, Courtesy of McDonnell Douglas Astronautics Company; 14, 14-15, © Michael Carroll; 16-17, © Mark Maxwell 1983; 17, 18, © Michael Carroll; 19, © Julian Baum 1989; 20-21, 21, © Rick Sternbach; 22, © Doug McLeod 1989; 22-23, Courtesy of Airstream, Inc.; 24-25, © Michael Carroll; 25, © Père Castor Flammarion; 26 (upper), NASA; 26 (lower), 26-27 (both), © John Cancalosi/Tom Stack and Associates.